# The Leghorn Chicken of the Past and Present
## A Short History of the Leghorn Fowl

by Frederick H. Ayres

**with an introduction by Jackson Chambers**

# Self Reliance Books

Get more historic titles on animal and stock breeding, gardening and old fashioned skills by visiting us at:

http://selfreliancebooks.blogspot.com/

# *Introduction*

I am pleased to present yet another title in the "Chicken Breeds" series.

This volume is entitled "The Leghorn: Past and Present". It was originally published in 1878 and contains a short history of the Leghorn.

The work is in the Public Domain and is re-printed here in accordance with Federal Laws.

Though this work is a century old it contains much information on poultry that is still pertinent today.

As with all reprinted books of this age that are intended to perfectly reproduce the original edition, considerable pains and effort had to be undertaken to correct fading and sometimes outright damage to existing proofs of this title. At times, this task is quite monumental, requiring an almost total "rebuilding" of some pages from digital proofs of multiple copies. Despite this, imperfections still sometimes exist in the final proof and may detract from the visual appearance of the text.

I hope you enjoy reading this book as much as I enjoyed making it available to readers again.

Jackson Chambers

Kellerstrass Farm

Arthur Oscar Schilling
1907

# PREFACE.

In writing of Leghorns, we are venturing upon untried ground, for, as in the second volume of this series, the Plymouth Rock, we are discussing a breed which has never been the subject of a book. The rapid sale of our previous books has assured us of the popularity of the project of publishing low-priced monographs on the various breeds and has led us to spare no pains to make this, the third of the series, worthy of the complimentary notices which the first two numbers have received.

How well we have succeeded, we leave it for our readers to decide.

# Leghorns.

THE Leghorn fowl is unique among poultry for various causes, but for none more than the fact that its birthplace is known, and that Leghorn fowls can to-day be procured in Italy, which in all features resemble the Leghorns found in this country.

Lying on the west coast of Italy, Leghorn is a natural port, and the stopping place of the vessels which ply between foreign ports and places on the Mediterranean. Leghorn has particularly close relations with America, as can be seen from the fact that its imports for the year 1872 amounted to $1,180,000, from the United States alone; and the tonnage of our vessels entering the port during that time was 1,079,455.

This place, then, was naturally the one from which the Italian contingent of poultry for the ever-ready American fanciers must be drawn. Outward-bound vessels almost always take on board at this port a stock of poultry and cattle for the homeward trip, and the fowls which survive the demands of the captain's table and the forecastle, were called—in default of a better name or in despair of pronouncing the Italian one, even if it was known—Leghorns. Concerning their first importation, however, there is some dispute. Doubtless, many reached our shores before they, came to the notice of anyone sufficiently posted on poultry matters to know their value and appreciate the importance of securing them. However that may be, the earliest record we find of their culture in America, is that given by Mr. O. H. Peck, of Franklin, Mass., who in a communication which appeared in the September, 1875, *Poultry World*, gives the following account of the Brown Leghorns.

"About forty years ago (*i. e.* 1835), Mr. N. P. Ward, of Fulton Street, New York city (the then celebrated cracker baker), received a few of these fowls, as a present, direct from Leghorn; this is, I think, the first record we have of them in America. The eggs from these fowls were distributed among his friends, one of whom was Mr. J. C. Thompson, of Tompkinsville, Staten Island, once an eminent poultry raiser, now deceased. Mr. Thompson wrote as follows:

"'I raised, from six eggs, five cocks and one pullet; the size of the comb and wattles of that lot, exceeded anything I have ever since seen. The length of the combs (actual measurement), was six inches. The comb extended so far out over the beak, that it was in the way of their picking up grains, and they were compelled to press the protruding comb on one side to get their bills to the ground.'

"On two or three occasions subsequent to this, Mr. Thompson obtained the fowls direct from Leghorn, once through his son-in-law, who was master of a vessel. They matured early, pullets laying at four months old; and Mr. Thompson was of the opinion that they oftentimes died from exhaustion, actually laying themselves to death. I once knew a party who had them and sold the eggs under under the name of 'Sicilian fowls.'"

From this account we get our first knowledge of Brown Leghorns in America, and also the information that they were sometimes called Sicilians, a fact which has an important bearing, as we shall show later, on the question of the purity of Rose-comb Brown Leghorns.

There has been an almost limitless amount of paper and ink lavished on the question of the first importation of Brown Leghorns, and I. K. Felch, in an article on Leghorns, in the February, 1873, *Poultry World*, gives the date of their first importation as 1855, probably a misprint for 1853, the date of the second importation to the yards of Mr. Roswell Brown, of Mystic River, Conn., as, in one of his books, he credits Mystic with the first importation. As we have already hinted, it is very probable that other Leghorns reached our shores between the supposed date of 1835 and the certified record of 1852, now in Mr. Brown's possession. Still 1852 is the earliest exact date, and may be looked upon as the initial year of Brown Leghorn breeding it this country.

About 1858 we had an importation of White Leghorns, known as the "Lord importation," and in 1863, the "Stetson" birds, which were much more nearly like a Standard White Leghorn than the first importation. The Lord fowls had white legs, but the Stetson birds were yellow-legged and had clean yellow beaks, with pure white plumage and good combs.

## Rose-Comb Leghorns.

From the time of the general introduction of Leghorns into America, the single comb was considered by almost all fanciers as the only one allowable. The reason for this decision is not far to seek; the majority of the birds purchased and bred were single-combed. The original owners of the imported fowls may have had rose-combed birds in their flocks (in fact we know this to be the case), but they sedulously kept them in the background. The reason for such an action being that buyers were accustomed to the single combs of Black Spanish fowls and by some occult method of reasoning had decided that a single comb was the characteristic of *all* the fowls which came from the shores of the Mediterranean. Whatever influence determined the matter, the single combs were successful and their rose-combed brethren went to pot with the greatest regularity.

## ROSE-COMB WHITE LEGHORNS.

Such was the state of affairs up to 1876, when **Mr. Charles F. Starr**, of New London, came across a rose-combed White Leghorn cock and secured him, as he states below. The fact that Mr. Starr had originated a breed of rose-combed White Leghorns came to our knowledge a short time since through a notice in the *American Poultry Yard*, and we at once wrote to Mr. Starr, for information on the subject. He responded promptly as follows:

" In the year 1876, on a farm in the town of Groton, I came across a *Rose-Comb White Leghorn Cock.* At least to all appearances he was such, for his plumage was clear and his legs and beak clear yellow, ear-lobes pure white and his style and carriage was that of a Leghorn. How he came there I do not know, as the other Leghorns on the place had single combs. I bought him and mated him with seven White Leghorn hens (Boardman Smith stock.) Out of about twenty chicks, hatched from this cross, about one-half had good rose combs. Taking seven of the pullets, I bred them back to the old cock with good success, and out of more than a hundred chicks hatched this last spring, only about twelve had single combs. That they will breed true there is no doubt, for as an *experiment,* I bred a cockerel from the first cross back to the old hen, and about half of the chicks from these had double combs. I would not be afraid to wager most any amount that the stock I have now will produce nine rose-comb chicks out of every ten hatched, and in another year I think that every chick from my stock will have good *rose-combs*.

"I have a beautiful flock of these birds now, and shall spare no pains or labor to bring them to perfection. They are much handsomer than the single-comb variety, and not liable to freeze their combs and are just as good layers. When these fowls are admitted in the Standard (and it is only a question of time), they will be in as great demand as the single-comb variety, and in northern climates greater, as the comb is not liable to freeze, and thus not only hurt their appearance, but stop their laying, for a time at least."

After receiving the above account from Mr. Starr, we took the pains to see the stock from which he obtained his rose-combed cock and assure ourselves that the birds were genuine, and that no cross-bred bird had been sold Mr. Starr, as a full-blooded Leghorn. The result of our investigations was perfectly convincing: not only were there no mongrels kept by the owners of the bird, but there were no white fowls kept in the neighboring yards.

This would seem proof enough of the existence of pure Rose-combed Leghorn stock, but the information we accidentally received from Mr. Reed Watson, the famous importer of Black Leghorns, was even more absolute. In October, 1878, Mr. Watson received an importation of Leghorns direct from Italy, that contained a *Rose-Comb White Leghorn* cock. Mr. Watson's order was for specimens of all the clearly marked varieties of Leghorns to be found in Italy, and he received Black, White, Brown and Spangled. Can anyone ask for more or better proof of the genuineness of these newly-found birds?

## Rose-Comb Brown Leghorns.

The agitation of the Rose-comb Brown Leghorn matter is one of such recent occurrence that we feel sure of the interest of all who breed Brown Leghorns — and their name is legion—in the history of this variety. Our own attention was first drawn to the matter by the statement of a correspondent of the *Poultry World*, who stated that he had endeavored to obtain a pen of Rose-combed Leghorns, by breeding a single-combed cock to a common red hen and that the progeny showed good rose-combs, but had acquired the sitting instinct from the mongrel dam. This account led the other breeders, who had tried similar experiments, to relate their experience. In almost all these cases the experiment was, like the one we have just cited, that of crossing pure Leghorn blood with impure or mongrel. A natural result of this repeated crossing and the publicity given to it, was that breeders everywhere were distrustful of anything which claimed to be a simon-pure Rose-comb Brown Leghorn, and having in nine out of ten cases never heard of such fowls as produced by breeding pure stock with pure stock, decided that no rose-combed fowls of Brown Leghorn plumage, could be genuine Leg. horns. It was at this point that we, having noticed in the *Poultry Yard,* an expression of the prevailing opinion over the signature of C. R. Harker, wrote our first letter to that paper on the subject. Our motive in this was to correct the impression that had gained such strong hold, by a simple statement of the facts that had come under our observation, which were these:

As we have already stated, the first importation of Brown Leghorns which has a recognized date was that to the yards of Mr. Roswell Brown, of Mystic River, Conn., and so to obtain a full account of the breed from its birth in New England, we had recourse to Mr. Brown, who enlightened us on a number of matters of interest. From his account, we learned that the birds of the 1852 and 1853 importations, alike, threw almost as many rose, as single-combed chicks, but that, as the single-comb alone was Standard, the rose-combed progeny was used for market purposes only and carefully culled out. For twenty-five years Mr. Brown has had the same stock and has almost entirely eradicated the rose-comb tendency, yet, he showed us a trio of fowls from his pens that had well developed rose-combs. With this stock as a text we ventured to differ with Mr. Harker, as all the readers of the *Poultry Yard* are aware. We were simply relating *facts* and must own we were rather staggered when so good a breeder as Mr. A. B. Campbell, of Norwich, Conn., came to Mr. Harker's side and announced that he (Mr. H.), had the best of the *argument*. However, that matter needs no farther ventilation at present and we need add but this hint, for those who are in doubt whether they have genuine Leghorns or no, that the distinctive characteristic of the Leghorn, rose or single comb, is the non-sitting qualification. If we have a fowl of Leghorn shape and markings, which has no desire to sit—"stick a pin here,"—it is a genuine bird and just what it purports to be.

## SICILIANS.

As we mentioned in a previous chapter, the fowls once known as Sicilians, are, from the accounts of the best authorities, almost identical with Brown Leghorns and differ from the Standard single-combed Brown Leghorn to about the same extent as do the rose-combed birds. The Sicilian fowls, had cup-shaped combs and were known to numerous breeders before Brown Leghorns had been introduced to any great extent. Among others, Mr H. H. Stoddard, editor of the *Poultry World* and *American Poultry Yard*, had a pair of them and found them very fine layers but rather hard to keep with the comparatively low fences which sufficed for the rest of his flock of a thousand.

The fact that these cup-combed fowls were found principally about the time of the first large importations and were identified and acknowledged as Leghorns, by competent

judges, goes to show that the Leghorn fowl as it came from
the Italian coast, was not only a single but cup-combed fowl,
and furnishes a plausible *argument* to add to our *facts*, in
favor of the existence of Rose-combed Brown Leghorns.

## BLACK LEGHORNS.

With the discussion of this breed we trust their origin-
ator in America, Mr. Reed Watson, simply premising that
we know Mr. Watson, and have heard, from time to time,
his experience in endeavoring to establish this variety on
an assured basis of clean breeding and Standard excellence.
His first importation, though they had been in this country
a year, is recorded in the *Poultry World* of October, 1872,
in these words:

"Mr. Reed Watson, of East Windsor Hill, Conn., has
some Leghorn fowls, direct from the vessel in which they
were imported from Italy. We lately spent a day, and
consider the day well spent, in visiting them, for such importations are rare.

"Mr. Watson's birds show the unmistakeable Leghorn
form, even to the details of comb and wattles, and are as
thoroughly non-sitters as any of our acclimated strains.
They are very vigorous and active. The original fowls,
three in number, imported a year ago are now (October '72),
surrounded by a well-grown and numerous family. The

old hens have proved themselves remarkably prolific layers, and the pullets of last April are following their example. A brood of a dozen chicks can be seen, hatched September 1st, from eggs laid by pullets of this stock, hatched after the middle of April last. That is, the pullets reproduced when less than four months old."

From this account we see that the Black variety showed itself in no way inferior to the better known Whites and Browns. Yet, Mr. Watson was not entirely satisfied and had recourse to other importations, of which notices occurred from time to time. After some years he secured a stock that bred to his satisfaction and in conversation, expressed his satisfaction with his success. In 1878, he published the following, which is the latest information on the subject:

. . . "Mine is the only stock of the kind on the western continent. I think this breed has accomplished more than any other in the same period, and is of more value than many others. For production of eggs they are unequalled, and I have hundreds of letters to that effect. The eggs are large and pure white, finding a ready sale, at rates above the market value. One of your wealthy citizens, a lady, whom I supplied regularly, previous to a visit to New York, dropped me a line to bring her a supply of eggs to take with her, as she never had been able to procure eggs of such fine quality elsewhere.

"A breeder says that he had one hen that laid over 300 eggs in a year. A nest was once found, that was stolen, in which a hen had laid a peck measure of eggs. They are absolutely non-sitters. Their skin is of a light yellow, which is a good thing for table purposes. They are very intelligent; knowing more in many instances than some breeders of poultry do.

. . . "A farm is the proper place for them (in fact for any breed), where they will largely support themselves by foraging. The chicks are smart enough to take care of themselves, dodging the hawks. To be sure, the first stock did not throw so true to feather as desirable, but there are breeds that have done worse. It was a source of pleasure to me, after getting three hens and a cock from Genoa, Italy, in July of 1876; and out of seventy-nine chicks in hatched August and September (of which I raised seventy-seven), to find that maturing, every one of them was *black* in feather, had correct combs, and laid when three and a half months old. I knew then that the stock was valuable." . . .

From this account, we learn all that is known of Black Leghorns, which have not as yet become sufficiently known or cultivated to enable us to give the general opinion of their merits. They are now however " Standard," and as Mr. Watson's flock has increased largely, we hope soon to see its progeny more widely known and esteemed. As yet, we have never seen them exhibited although we have attended a great many poultry shows and examined all the stock carefully.

## DOMINIQUE LEGHORNS,

end the list of the Leghorn family and have the character-istic merits of their class. In number, they are deficient however, owing probably to their comparatively recent admission to the Standard and the small amount of " push " their advocates have employed to make them known. Like all the others of this family, they are splendid foragers and well adapted to the farmers' needs ; and in sharing the slatey-blue color of the Domin-ique and Plymouth Rock, they gain an additional claim on those who have pasture and wood-lands, where feathered stock can be kept to advantage, provided it is proof against dirt and discoloration. We know of no reason why this variety should not become as popular and widely cultivated as the Brown or White.

## LEGHORN PRECOCITY.

Under this head we shall give brief extracts from the records of different breeders, as they have appeared in different publications from time to time, or have come to our knowledge from the accounts of our acquaintances. In the *Poultry Yard*, appeared recently, the following:

. . . "Last year I thought I should like to try the

Brown Leghorns, having read so much about them in the
*Poultry World.* I looked through the advertisements for
the nearest breeder, and sent for two sittings of eggs. On
the 29th of August, 1877, I found to my surprise, twelve
fine Brown Leghorn chicks, one day before I expected them:
I raised all but one; four cockerels and seven pullets.
Sold two pair, hence had five pullets, which commenced
laying when a little more than four months old. I was
somewhat disappointed by the small eggs, so determined
not to keep them, sold them at a sacrifice. However, they
came into good hands, the size of their eggs became larger
as the pullets grew stronger, and I would have paid double
the amount received for them to get them into my hands
again." . .

In an early number of the *Poultry World,* at the time
when the fight over ear-lobes and other points was so hot,
we find the following, from a breeder who claims to have
bred Leghorns since 1853, a period of twenty-one years.

. . . "I have kept strict account with my fowls from
the beginning; and can, by comparing figures, see no differ-
ence in the time of their maturing and commencing to lay.
Some commence when three months and a half old, and
none commence older than five months. I could give figures in cases where I have 'timed'— if you please—twenty-five pullets nearly every year, for the last seven years,
and a less number of them for the last twenty years, were
it necessary.

"The best I have ever done, was the past year. I had a

large number hatched out the 9th of August and selected twenty-five pullets, to whom I gave an extra run and moderate feed for five months, when they commenced laying the 9th of January. These twenty-five hens laid, up to the 9th of August (a year from the date of hatching), 3,750 eggs, or 150 each; their average weight is five and one-half pounds, and they will lay before the next 9th of August, 240 eggs more." . .

On this topic, which is of such great interest to all breeders, Mr. C. R. Harker, the noted professional breeder of Brown Leghorns, in response to request for an account of his experience in this particular, writes as follows:

"I have had two Brown Leghorn pullets, hatched March 1st, lay: one June 10th, and the other June 15th, or at about three and a half months old. But this was where their growth had been forced ahead by stimulating food, and tender treatment. A Brown Leghorn pullet does *well*, if, with ordinary care, she lays at five months, but I never owned one, hatched before June 1st, that did not do her full share toward filling the egg basket, before the ensuing Thanksgiving. A peculiarity with a Brown Leghorn is its *capacity* for being *forced* along to maturity, by high feeding and good care. Given these, and a Brown Leghorn pullet approaches puberty, with that quickly developing voluptuousness which characterizes the growth of the maiden of its own sunny Italy. As to cockerels, I have no doubt that an unusually smart one can be, under proper conditions, the father of a numerous family at the ripe age of three months, at least I have seen cockerels at that age who seemed willing and able to do their full share toward bringing about such a result. You have my best wishes as to the success of your book, and I remain,

                        Very truly yours,

                                C. R. HARKER."

We might draw from the publications we have already cited, almost innumerable letters to the purpose, but could gain nothing by reiteration. In truth, it would be hard to present the case more forcibly than these accounts from practical breeders of reputation and long experience state it.

## LEGHORNS AS EGG PRODUCERS.

Though everyone who has even the merest smattering of poultry knowledge has heard of some of the astounding feats of this breed, and seen occasional accounts of their wonderful records, we cannot assert that every flock or every member of a flock will surpass any single specimen of any other breed, but that, taken as a whole, they are incomparable. With a view of showing this more clearly, we shall give some of the most extraordinary of the many egg-records which have appeared in the last six years.

Some years since, in speaking of this breed, I. K. Felch says: "These fowls as egg-producers, in their original per-

fection were truly marvelous. I have known of a hen of the last importation, laying 159 days in succession, and have the assertion of a friend, that one laid 275 eggs in one year; but the largest number of which I know personally, and which I deem very extraordinary, was 250. An average, in my experience has been from 175 to 200 eggs. With good care, 200 eggs need not be despaired of."

In 1872, Mr. Lynde, of Marlboro, Ohio, gave the results of a series of experiments instituted for the purpose of ascertaining the cost of keeping and the fecundity of different varieties as follows:

" The fowls embraced in the following experiment were hatched between the 25th of February and the 1st of March. When they were six months old, I put ten pullets of each breed in yards forty feet square, each yard containing a small warm house. These fowls were fed corn the first day, oats the second, and wheat screenings the third, then corn, then oats and then screenings, and so on for the entire time. The feed was put in boxes, always putting in enough so there would be some left at the end of the day. They were also given a little fresh meat as often as three times a week, plenty of water, and burnt bones. The first month (September), the

|  |  | Corn. Qts. | Oats. Qts. | Screenings. Qts. | Eggs. Laid. |
|---|---|---|---|---|---|
| Brahmas, Dark, | ate ..... | 25 | 18 | 21 | 127 |
| Cochins, Buff | " ..... | 28 | 22 | 19½ | 97 |
| Dorkings, Gray | " ..... | 21 | 17½ | 16 | 101 |
| Houdans, | " ..... | 15 | 14 | 14½ | 142 |
| Leghorns, | " ..... | 13 | 12½ | 16 | 161 |

In the entire time (six months), the

|  | Corn Qts. | Oats Qts. | Screenings Qts. | Eggs Laid. |
|---|---|---|---|---|
| Brahmas, ate............ | 142 | 108 | 119½ | 605 |
| Cochins,   "    ............ | 160 | 132 | 114 | 591 |
| Dorkings, "   ............ | 118½ | 100 | 91 | 524 |
| Houdans,  "   ... ........ | 93 | 61 | 60¼ | 783 |
| Leghorns "   .......... | 74½ | 77 | 80 | 807 |

Mr. Lynde also states the cost of the food supplied each variety:

| | |
|---|---|
| Brahmas............................. ..............$4 90 |
| Cochins............................................. 5 36 |
| Dorkings............................................ 4 45 |
| Houdans........................... ................. 3 34 |
| Leghorns........................................... 2 97 |

From this we see that the Leghorns not only laid a greater number of eggs, but did so at less cost to their owner. It has become so universally acknowledged that Leghorns are the most prolific of all fowls, that it hardly needs this detailed statement of their recorded excellence to enable them to hold their enviable reputation. Still the figures are valuable as a proof against the aspersions of cavilers of any sort.

In singular contrast to the details given above is the item which has travelled the rounds of the country press and breaks out from time to time with renewed vigor. We quote it entire as a gem worth preserving in the annals of poultry-*pseudo-science.*

"It has been ascertained that the ovarium of a fowl is composed of 600 ovules or eggs; therefore a hen, during the whole of her life, cannot possibly lay more than 600 eggs, which in natural course, are distributed over nine years, in the following proportion:

| | | | | | |
|---|---|---|---|---|---|
| First year after birth | | | 16 | to | 20 |
| Second | " | " | 100 | to | 120 |
| Third | " | " | 120 | to | 135 |
| Fourth | " | " | 100 | to | 115 |
| Fifth | " | " | 60 | to | 80 |
| Sixth | " | " | 50 | to | 60 |
| Seventh | " | " | 35 | to | 40 |
| Eighth | " | " | 15 | to | 20 |
| Ninth | " | " | 1 | to | 10 |

It follows that would not be profitable to keep hens after their fourth year, as their produce will not pay for their keeping, except when they are of a valuable breed."

Rather a striking contrast, is it not, to the production of Mr. Lynde's Leghorns? By computation based on *science* they laid somewhere in the neighborhood of 50 eggs each, or if we take the very best figures about 60, in the period between their tenth and eighteenth months. Yet Mr. Lynde gathered 807 eggs from his ten fowls in that time. But perhaps science was not acquainted with Leghorns.

We will close this record with one which was sufficiently extended to enable us to gain very valuable ideas as to the power of this breed to sustain a high average through the year, and also on the average fertility of a large flock. The record is that of Mr. A. J. Tuck, and is as follows:

" A neighbor of mine has a flock of White Leghorn hens, concerning which he has furnished me the following particulars. He commenced his daily record of eggs on the tenth of last December, with three eggs, and up to the tenth of August—a period of just seven months—he had received from his flock 4,279 eggs. At the beginning of his record, his flock consisted of thirty hens and pullets, some of which did not commence laying until the last of March. During this time he has had two die, so it left him twenty-eight all told. Here we have an average of, in round numbers, 150 eggs per hen, in but little over half a year."

When we consider that in this flock of thirty all told, a number were pullets, which did not commence to lay till some time after the record was begun, it is probable that we should find—with a more careful record, crediting the pullets as members of the flock only as they began to lay— that the average was more nearly 175 than 150.

# THE

# PLYMOUTH ROCK,

AS THE

## Fowl for General Use;

WITH

### Rules for Mating and Breeding According to Nature.

## PRICE 15 CENTS.

UNIFORM WITH

# THE GAME FOWL,

AND

# THE LEGHORN.

Address,

## F. H. AYRES,

### MYSTIC RIVER, CONN.

# Egg Record Cards.

## A Daily Calendar,

GIVING

### A PLACE FOR A COMPLETE RECORD

— OF —

No. of Fowls;

The Variety;

Eggs laid each day;

Eggs sold;

Fowls sold;

Chicks sold;

Expenses, and Balance for the Month.

☞ Printed on fine Tinted Bristol Board, each month on a separate Card.

## PRICE 15 CENTS PER YEARLY SET.

*By mail, prepaid, to any address.*

Address,

### Mercantile Printing House,

HARTFORD, CONN.

www.ingramcontent.com/pod-product-compliance
Lightning Source LLC
Chambersburg PA
CBHW060007230526
45472CB00008B/1988